Solar Heating for the Home

Also by Graham M. Hunter (as G. St. John Murray)
A Practical Guide to Retirement

SOLAR HEATING FOR THE HOME

Graham M. Hunter

DAVID & CHARLES
NEWTON ABBOT   LONDON   NORTH POMFRET (Vt)

697.78
H945a

British Library Cataloguing in Publication Data
Hunter, Graham Martin
    Solar heating for the home
    1. Solar heating – Amateurs' manuals
    2. Dwellings – Heating and ventilation – Amateurs'
    manuals
    I. Title
    697'.78   TH7413

    ISBN 0-7153-7726-4

First published 1979
Second impression 1980

Printed in Great Britain
by Biddles Ltd, Guildford
for David & Charles (Publishers) Limited
Brunel House Newton Abbot Devon

Published in the United States of America
by David & Charles Inc
North Pomfret Vermont 05053 USA

Contents

Foreword
by Leslie Smith

Any conscientious doctor with a large practice has more than enough to keep him busy. Opportunities for leisure pursuits are few enough, and it takes a man of unusual enterprise to develop new and varied talents.

Graham Hunter is such a man. Thus he has written books and articles on many different subjects of social importance. And over the years he has repeatedly given his expertise to BBC documentary broadcasts presented by me on topics ranging from teenage problems to care for the elderly.

He is a naturally practical person. But practical people often find it difficult to accept that the rest of us can feel clumsy and foolish when faced with wires, pipes or tanks. Everyone reading this book, however, will find it pleasantly logical and easy to understand.

Preface

This book is an introduction to the concept of solar heating. In it I have sought to describe the various ways in which this form of heating can be used, and to give practical advice for anyone wishing to install solar heating in a house.

Many different types of solar-heating apparatus are on sale at widely differing prices. Imposing claims are made for their performance. Often they will only yield the heat claimed if water is used continuously, which is seldom the case in a normal home. Just what can be expected from different installations is discussed here, as is the actual installation of solar-heating apparatus.

Each chapter is complete in itself, with any necessary reminders to the reader of earlier information on aspects of planning or installing a solar-heating system.

ACKNOWLEDGEMENTS

I should like to thank Leslie Smith of the BBC for writing the Foreword, and Peter Smith of Thermal Systems Ltd for his helpful advice and for providing some of the illustrations. I should also like to thank my secretary, Mrs K. Davey, for her patience and skill in typing the manuscript.

Graham M. Hunter

1 What is Solar Heating and How Does it Work?

Advertisements for solar-heating systems in newspapers and periodicals have raised a number of questions among those of us who would like to reduce our spending on fuel and save energy resources, while keeping a good supply of hot water in our homes. What is this new form of energy? Is it really free and is it suitable for use in my home? What about my existing boiler? Which system should I choose and is the whole project going to be worthwhile?

Filling in the coupons found in various newspapers brings details of many types of installations at various prices: But unless the basic principles are understood it is virtually impossible to compare one installation with another, let alone make a sensible decision about installing a system in your own home.

In the chapter that follows an attempt will be made to help explain the various systems so that you can make up your own mind about how they match what you need, and at the same time avoid the many pitfalls that can trap the unwary.

There is nothing new about using the best from the sun. We use it every time we sunbathe, when we plant our peaches against a south-facing wall, or grow plants in greenhouses, cold frames and cloches. We also use it to evaporate water, whether commercially in the great salt-pans of the world or domestically to dry the weekly wash.

What is new about solar heating is that we are now able to trap the heat with a specially designed unit, the solar panel, and transfer it to our domestic water supply to save heating some of that water by other means.

It is an amazing experience, even for experienced plumbers, to plunge a hand into the outflow from a solar panel and feel the heat in the water. Sometimes it is too hot to leave the hand in it. This is the heat that is available free and for a large part of the year. How can we trap it and put it to use?

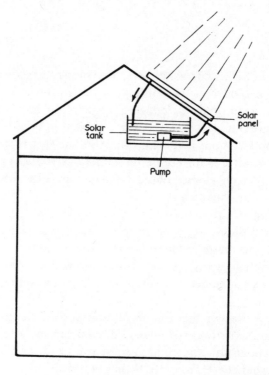

Fig 1

Here we see a simple diagram of a house with a solar panel on its roof. The warmth from the sun is attracted to the blackened surface of the panel which becomes warm in consequence. The water from the solar tank in the attic is pumped through the solar panel, taking up this warmth.

On a sunny day the water in the solar tank becomes surprisingly warm and can easily reach 50°C (120°F), which is hot enough for most domestic purposes. It is using this heat, raising the efficiency of the system and incorporating it in the domestic hot water supply, that create the technical problems that are now being overcome.

The great drawback to solar heating is the difficulty of storing the heat acquired. Once the sun has heated the water, somehow the tank must be insulated so that the minimal amount of heat is lost. Some engineers have experimented by having a huge water tank in the basement of a building and trying to store the heat in this. The practical problems are formidable as many thousands of gallons of water are required, and a huge cost is involved. As a compromise the heat is stored in a smaller water tank in the roof of the house gauged more to the immediate needs of the occupants.

There is another way of using the sun's heat as energy, and that is by means of silicon solar-electric cells. These convert the sun's energy into electricity. Silicon cells are exceedingly expensive and have only a small output, but because they have no moving parts they need a minimum of maintenance and have a long life. The present high cost of refining the silicon from sand may sometime be reduced, and in this case their value for domestic use would need to be reconsidered.

The latest scheme from the United States uses this system in conjunction with large mirrors to reflect extra sun onto the cells. I mention this as it is not unusual for people (including those who should know better), to confuse this form of solar heating with the solar heating of water, which is the main subject of this book.

The Solar Energy Available

The amount of solar energy available is considerable: for the technically minded, it amounts to 2½ kilowatt hours of radiant energy for each square metre of the earth's surface per day. The enormous potential of all this heat is not really appreciated. It has been calculated that the solar energy falling on an area only 60 miles square in a favourable location is greater than the entire world's energy demand.

Despite its high cloud cover, the United Kingdom receives

11

more than half the solar energy reaching Australia per unit area. Because of the large water-droplet size a large proportion of radiation penetrates the clouds.

Most places in Britain get about 1,500 hours of sunshine a year, and seaside resorts proudly quote this in their holiday brochures. Below are listed the hours of bright sunlight for various places in the country. When the sun does shine it is not much less hot than in the tropics, as anyone who works in a glass-sided office can tell you.

London (Kew)	1486	York	1310
Croydon	1528	Hull	1330
Margate	1771	Scarborough	1391
Southampton	1639	Bradford	1217
Plymouth	1674	Sheffield	1252
Birmingham	1302	Aberystwyth	1484
Oxford	1482	Haverfordwest	1606
Cheltenham	1442	Cardiff	1566
Norwich	1572	Shrewsbury	1339
Yarmouth	1616	Aberdeen	1317
Luton	1486	Edinburgh	1385

These figures show that there is a tremendous amount of untapped energy available, and much incentive for all the research into how to harness it that is going ahead, particularly in the United States.

A United States National Science Foundation report predicts that solar energy could provide 35 per cent of the heating requirements of the United States and 20 per cent of its electricity. President Carter himself has allocated a large sum per year to be spent on solar-heating research. He has installed solar heating in the White House and has allowed large tax rebates to people doing so in their own homes.

A similar study by the Australian Academy of Science calculates that solar energy could provide 40 per cent of Australia's low-grade heat.

The Japanese, for whom lack of fuel is one of the main industrial problems, are turning their minds to tapping solar heat. Already millions of units are being supplied to their houses and they are beginning to enter the export market. The South Africans, Israelis and French all have large solar-research programmes.

In Britain, research into solar heating and other forms of energy conservation is going ahead at the National Centre for Alternative Technology at Machynlleth in Wales. Here various types of solar-heating panels can be seen in action, together with windmills, methane-gas generators and many other types of energy-saving machinery.

Solar Heating in the Home

Solar heating is the simplest of these forms of energy saving and the one most easily adapted for general use. Indeed, on the domestic front all sorts of claims abound. Some advertisers claim up to 40 per cent of one's fuel requirements can be met by solar heating; others that sixteen black solar panels can raise the outside temperature of a swimming pool from 35° to 57°F (2° to 14°C) even on a dull day. It is said that even in winter you can raise the temperature of the water in your solar tank to give up to 15 to 20 per cent of your needs. How can this be achieved?

There are many solar-heating installations already working today. In Texas there are thirteen coin-operated laundrettes powered by the sun, while in England, Pontin's Holiday Camps use solar power to heat many of their swimming pools. So successful is the venture that they are opening up a new factory in Cornwall to produce solar panels. Other firms are increasing their output of solar equipment, and some giant British companies are preparing to move in once the idea really catches on.

Recent reports by the Select Committee on Science and Technology recommend that the installation in domestic premises of approved solar water-heating systems should qualify for a grant of 50 per cent of the capital and installation costs subject to a maximum of £400. If this report is ever implemented there will be a tremendous surge in the production of solar equipment and a consequent reduction in its capital cost.

Selecting Your System

There are many small firms, now over sixty in Britain, offering to install solar heating in private houses and the information in this book should give some guidance as to which, if any, system to select. Not everyone has been happy with commercially installed systems and they do tend to be expensive. Do-it-yourself systems are also available. These allow a certain amount of latitude in the way they can be set up and one can always enlist the help of a local plumber to make the actual installation. This can work out very much cheaper and improvements can easily be made to the system as you go along.

The Basic Unit

What is it that collects this solar heat and how does it work?

The basic unit for collecting solar heat is the solar panel. In its very simplest form this can be a blackened sheet of corrugated iron set up at an angle facing the sun. A thin stream of water flows down the sheet to be collected, surprisingly warm, by a gutter at the bottom. Simple though this method is, it has been used to heat some swimming pools, and the design has even been simplified so that water is merely pumped over a blackened area of the swimming pool surround and allowed to run back into the pool itself.

The above methods are crude, inefficient, and open to contamination from dust and dirt. The modern solar panel is an enclosed system which can be of many designs, as described in the next chapter; it allows water to circulate between two blackened areas, becoming warm in the process.

Much has been done to raise the efficiency of this system, partly by varying the composition of the solar panel and partly by trying to mount it in the best possible position.

So far we have seen that solar energy is available for

domestic use. There are various types of apparatus which can be installed to make use of it. The potential saving in conventional fuel is considerable and there is also the added advantage of having many gallons of partly warmed water available in the house which can be very useful when hot water is particularly in demand, as on washdays or when several people are taking successive baths.

2 Types of Solar Heating

The simplest form of solar heating is achieved by putting a water tank on the roof. This is quite common in some tropical countries where the flat roofs of the houses lend themselves to this purpose and the water is heated surprisingly well.

The next simplest method, applicable to more temperate zones, is to trickle the water down an inclined sheet of corrugated iron, as described in Chapter 1. But we have seen that to heat a domestic water supply a more elaborate system is needed though the same principles apply.

Thermosyphon System

The next step forward is to pass cold water through a solar panel mounted and angled to face the sun and it emerges warmed at the top (see Fig 1). As warm water rises anyway it might be thought possible to do without a pump. This can be done using what is called the thermosyphon system. Figure 2 shows a solar panel from which the warm water rises slowly to the tank at the top. It is replaced by the denser cold water from the bottom of the tank which enters the solar panel at the bottom. This system can be run with one or more solar panels according to requirements, but the principle remains the same. It is the simplest possible method for a house, and it can be put together by a handyman at very little cost. Though not very efficient, it can be used under certain circumstances where higher temperature and efficiency are not so important.

The main drawback to this system is that the flow is so sluggish that most of the efficiency is lost. Heat would also be lost by the solar panel radiating its heat away every time the sun passed behind a cloud.

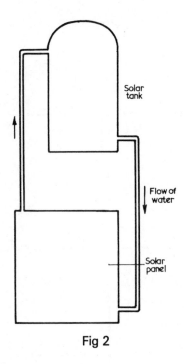

Solar
tank

Flow of
water

Solar
panel

Fig 2

Pumped System

This problem is overcome in a pumped system. The incorporation of temperature-sensing devices at the solar-panel outflow and in the water tank can shut off the pump when there is no useful heat to be gained from the panel, switching it on again when warmer conditions prevail.

In the normal pumped system the warmed water in the solar-heated tank circulates through the solar panel again and again until it has become as warm as it possibly can. Water from the solar-heated tank can be used in two ways. The first is to draw it off directly with a tap leading to the bathroom or kitchen, so that on favourable days it can be used as the sole supply of warm water. This can be satisfactory so long as the distance from the tank to the tap is short (Fig 3). The second, and rather more usual, method is to use the warmed water to feed the cold-water inlet of the normal heating system. This is where the saving in fuel comes, as you are feeding your boiler

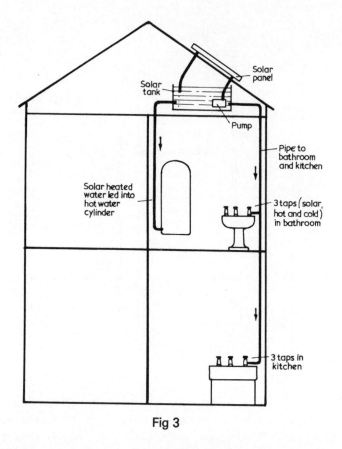

Fig 3

with pre-warmed water instead of with water at the usual
mains temperature of 10°C (50°F).

For most of the year and, surprisingly, even on certain days
in midwinter, a substantial amount of warmth can be collected
by this means. It is to improve the efficiency and consequently
increase the amount of fuel saved that many variations on the
above method have been devised. Understanding some of these
methods enables you to compare what various systems offer
and to decide which one, if any, is most suitable for your own
particular house.

When sun falls on a surface, part of it bounces back as a
reflection, leaving only the rest to be absorbed as heat. As black
colours cause least reflection, all solar panels are painted matt

18

black; and as the heat lost by reflection is greatest when the angle to the sun is large, solar panels should be sited so that they face the sun as much as is possible.

The Greenhouse Effect

Some manufacturers cover the solar panel with a sheet of glass—thus some of the rays which may be reflected off the solar panel hit the glass and bounce back again. This is called the greenhouse effect, and is why greenhouses become quite warm on only moderately warm days.

It works in this way. The sun's radiation is of a short wavelength; it passes through the glass and is absorbed by the black panels. Radiation from the panels is of long wavelength which does not pass through the glass so easily. Therefore the heat

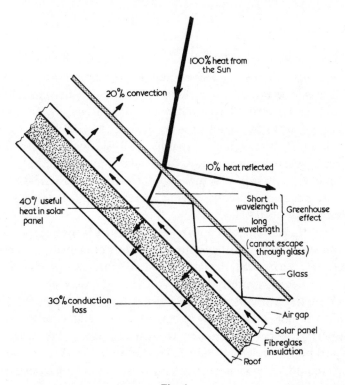

Fig 4

energy is trapped. This effect can be increased further by special selective coatings, though this involves considerably more expense. Copper oxide is commonly used in special paints to make a surface more absorbent to heat.

Some manufacturers try to improve upon this by having a double or even a triple layer of glass over the solar panel. There is a limit to the effectiveness of this method as each layer of glass reflects its own amount of light and its efficiency becomes much less once it gets dirty. Layers of glass also increase the cost of the solar panel as well as its weight, both of which are important factors.

Some heat can go right through the solar panel and escape through the back; to guard against this each panel must lie on an adequate bed of insulation, usually glass wool.

Siting the Solar Panel

It is obvious that to collect most sun the panels will have to face south, but as the sun's angle changes continuously as the day advances it is impossible to have the solar panel continuously facing it. It would have to be mounted upon a constantly moving platform, and as the height of the sun varies continuously through the seasons, the platform would also have to rotate up-and-down to ensure that the sunlight fell upon it at right angles. A moving platform is impracticable, so it can be seen that the siting of the panel is at best a compromise.

Nevertheless, the best possible compromise is most important —especially as some of the sun's rays do not fall direct, being deflected by passing through cloud, so that the panel misses them however carefully angled to the sun it may be.

Often the available sites are limited, usually to the roof or a wall, though occasionally a panel can be set on timber supports in the garden or over the garage. The situation should be unshaded, at least from the hours of 10am to 5pm, and should face as near due south as possible between the limits of south-west and south-east. These limits have sometimes been extended: useful heat can be obtained from an installation on a west wall.

An unfavourable situation would require a larger solar-panel area to compensate for the diminished sunshine.

The angle tilt is less critical, and can be from 20° to 70° from the horizontal. A 30° tilt gives the maximum total radiation over the year, but 50° enables better collection efficiency in spring and autumn.

As most people fix their solar panel on the roof, the angle of tilt is already determined by the slope of the roof. Most modern roofs are designed to have pitch of 40° to the horizontal. Where one is able to choose the angle of pitch it would be most economical to make it the same as the degree of longitude at the site plus an additional 10° to 13° to make the panel more efficient in the winter when the sun does not rise so high.

Vertical collectors, sited on a wall, can be used, but must be of larger area to make up for their decreased efficiency.

Inside the Panels

Quite apart from solar panels being covered by various layers of glass, their internal arrangements can differ considerably. Some panels consist of two thin sheets of plastic spaced a short distance apart, between which a thin layer of water flows slowly upwards collecting heat as it goes (Fig 5).

Some collectors consist of a coiled length of plastic piping arranged like a catherine wheel, with the water entering the

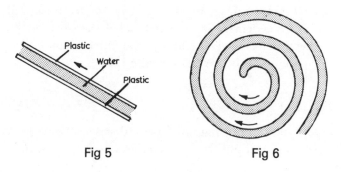

Fig 5   Fig 6

middle and gradually flowing round and round till it emerges at the out coil suitably warmed (Fig 6).

Some panels are extruded, so that there is a spongy network of passages: the water is pumped through the entire thickness of the panel, emerging eventually at the top (Fig 7). Other panels consist of metal tubes going back and forth (Fig 8), while others have channels pressed out of metal sheeting.

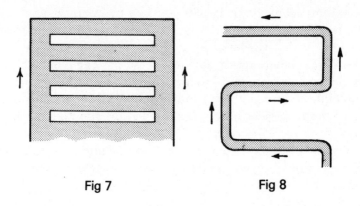

Fig 7   Fig 8

There are many variations of these methods. Some panels contain copper or steel pipes bonded to the outer cover. These pipes need to be close together to obtain the most heat, and to be near to the glass surface if there is one; they contain only a small volume of water.

The pipes are usually backed by a layer of aluminium foil to reflect back heat, and the whole panel rests on a layer of insulation to prevent heat loss from the back.

Representatives from firms specialising in solar heating say that sun-catching equipment is unlikely to be improved, though a more expensive solar panel called the Miramit has recently been developed in Israel. Doubtless many more ingenious panels will be invented in the future.

Choosing Solar Panels

Most solar panels in normal use have efficiencies between 30 and 40 per cent and only small increases in efficiency can be achieved by added sophistication. Selective absorbers, for example, offer some advantage over the normal surface, but the added cost is high in comparison to the benefit received.

Actual efficiency in a mathematical sense depends upon how the panel is used, and if relatively cool water is regularly drawn off in quantity the efficiency will be highest; but if warmer water is drawn off in smaller quantities intermittently the efficiency will be less, though this is what is usually needed in the domestic situation.

In practice it is desirable to aim for a maximum water temperature of 50°C (120°F), which is hot enough for most domestic purposes without further heating.

How then can we weigh up the advantages and disadvantages of the solar panels available? Mathematically good examples of each type of panel all have similar steady-state efficiencies.

Plastic types have the advantages of being low-weight, low-cost and chemically inert. They must be manufactured to a high standard, of tough plastic with high impact strength and a melting point that is reasonably high (430°F, 220°C). They must also be stabilised against degradation by ultra-violet radiation.

The plastic inserts of solar panels can crack and cause leaks. Sometimes the inserts are damaged while they are being fitted, and some damage can result from them being hit by falling debris. Occasionally cracks are caused by a design fault. The cost of replacing the plastic insert is very much less than the cost of replacing the whole solar panel. It is, therefore, worthwhile enquiring whether this is possible on the make of panel you intend to purchase.

Panels containing metal components, particularly aluminium, are liable to corrosion and this can be a very serious problem. Because of this they may not be suitable for use in the methods so far described, in which the actual water from the solar tank flows through the panel; they are reserved for the indirect method of heating (shown in Fig 11). In the indirect method the problems of corrosion are overcome by filling the solar panel and the heat-exchange coil with a corrosion inhibitor and antifreeze. Ethylene glycol is usually recommended as the antifreeze and sodium nitrite or benzolate as the corrosion inhibitor. This latter is often incorporated in the antifreeze, in which case it is called 'inhibited' ethylene glycol.

This system involves the use of a special header tank, or sometimes the system is permanently sealed. It is not suitable for installation by amateurs, and even if installed by experts the householder must be aware that there must be no leakage from the solar panel heat-exchange unit into the water of the solar tank. A special seamless piping must be used for the heat exchanger and the Water Authority would need to be informed.

In spite of this, many installations do run along these lines and are apparently satisfactory. Nevertheless, there is always at least the theoretical risk of the water supply becoming contaminated should a leak occur. The dangers are lessened by the fact that any leak would be small and at least it would be into the hot-water system from which water would not normally be drunk.

The quantities of antifreeze and corrosion inhibitor used are determined by the manufacturers and are more than sufficient to allow for any extremes of temperatures that may occur in their use.

Some manufacturers of plastic solar panels say that antifreeze or frost precautions are unnecessary, as their panels expand sufficiently to allow for ice formation. This claim must be treated with reserve. The capacity of plastic to expand can vary with its age, the surrounding temperature, and the long-term effects of ultraviolet light.

In the indirect method some efficiency is lost as heat can only be transferred from the panel by means of a coil.

Another consideration when selecting solar panels is how much water (or antifreeze) they contain. The greater the volume, the greater is the panel's thermal capacity; but high thermal capacity can be a disadvantage in certain conditions, as short periods of sunshine cannot be turned to good use because of the long heating time of the solar panel.

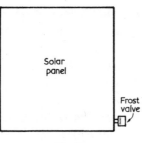

Fig 9

How complicated is the panel? Sophisticated panels are needed most where a high temperature is wanted. Simple panels are sufficient for low-grade installations for swimming pools. Some of the more expensive methods are heavier and more elaborate while not necessarily collecting more heat.

Precautions must be taken against freezing. Usually with the direct method, as seen in Fig 10, the panel is designed to be self-draining and incorporates a frost valve (Fig 9), which opens when the temperature falls near freezing and allows the panel to drain.

The indirect system, as shown in Fig 11 and described above, is usually filled with antifreeze so that there is no danger of frost damage.

Panel to Tank

How can one use the heat that has been collected? As the solar-heated water leaves the solar panel there are two ways in which it can be used. Either it can be run direct to the hot-water cylinder or it can go into a solar tank.

It has been found that to lead warmed water into the hot-water cylinder direct from the solar panel is unsatisfactory, due to a number of factors, including the width of pipe necessary to maintain a good flow, the distance of the solar panel from the cylinder and the intermittent nature of the production of solar heat. Nevertheless, some systems are planned along these lines and great care must be taken in evaluating them before agreeing to have them installed.

Although it makes the system appear more complicated there is no escaping the fact that the only satisfactory way of transferring solar heat to the hot-water cylinder is to route it through a preheat or solar-heated tank. This is the tank that has been described in previous diagrams along with the solar panel; it is an additional tank introduced into the house, quite separate from the normal cold-water cistern already installed. It will be of some 40 to 50 gallon capacity, well insulated and installed near to the solar panel.

There are two methods of transferring heat from the solar panel to the solar tank—the direct method and the indirect method.

The direct method. This is the simplest, the least expensive and the most efficient method. It simply involves leading the outflow pipe from the solar panel directly into the solar tank (Fig 10). Also in the solar tank is a submersible pump, which will pump the water round through the solar panel until it has become as warm as circumstances allow. Thermostatic sensors, previously mentioned, stop the circulation when there is no more heat gain to be obtained.

The indirect method. In the second or indirect method (see Fig 11) the solar-heated water emerging from the panel passes down into the solar tank, where it enters a coiled metal pipe. No mixing takes place between the water from the solar panel and the solar water tank but heat is exchanged in the coil and passes into the solar water tank. The now-cooled contents of

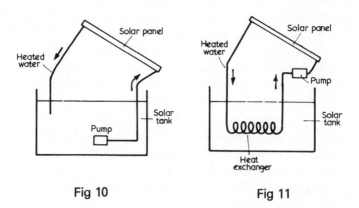

Fig 10   Fig 11

the coil are pumped up again through the solar panel to be rewarmed, and the thermostatic sensors control this circulation as in the direct method.

Whereas the first method is simpler and cheaper to operate, the second method has the advantage that antifreeze can be added to the solar-panel circuit; indeed, this circuit can be sealed permanently. Against that, the system is less efficient and the coil can only transfer heat when it is hotter than the tank water so that small temperature gains cannot be picked up.

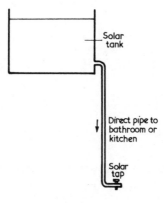

Fig 12

Tank to Tap

There are two main methods of taking the warm water from the solar tank to where it is used. One is to have a special pipe going to a solar tap in the kitchen or bathroom, or both, and if this is practicable there are great savings in fuel—in summertime the water is often hot enough to use without further heating. The second method is to lead the warmed solar water into the main household plumbing system. This is either done by leading the actual solar-heated water into the base of the hot-water cylinder (Fig 13), or by taking the heat to the hot-water cylinder by means of a heat-exchange coil (see Fig 14).

27

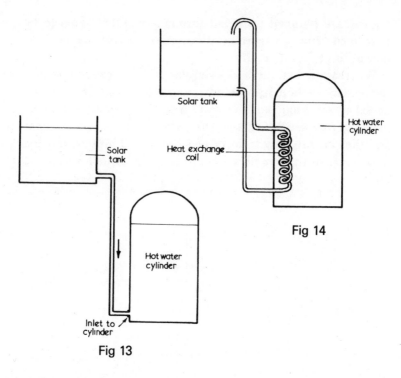

Solar tank

Solar tank

Hot water cylinder

Heat exchange coil

Hot water cylinder

Fig 14

Inlet to cylinder

Fig 13

To understand how this is actually accomplished in practice one must first appreciate the set-up of the plumbing in the normal home.

Figure 15 shows a simplified version of the normal water system of a house. Here we see water from the mains being directed via a ballvalve into the cold-water cistern usually situated in the attic. One pipe leaves this cistern to feed the cold taps, while a second pipe goes down to the hot-water cylinder to keep it full. This cylinder is heated either from a boiler by means of a heated coil or by an electric immersion heater. Hot water leaves this cylinder when drawn off to feed the hot taps, or when it feeds through a vent going to the cold-water cistern (or to its own header tank) to allow for expansion and contraction in the cylinder.

It is to feed this system with solar-heated, rather than boiler-heated, water that the various means have been devised. The problem is how to do it safely, economically and easily.

28

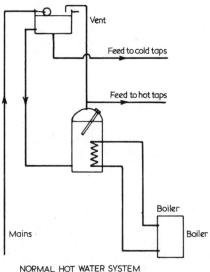

Vent

Feed to cold taps

Feed to hot taps

Boiler

Boiler

Mains

NORMAL HOT WATER SYSTEM

Fig 15

The direct system. In the direct system described in Fig 13, you — or your plumber — cut off the cold-water feed to the hot-water cylinder from the cold-water tank, and feed the hot-water cylinder with warm water from the solar-heated tank. Thus, we see that water from the mains goes to both the solar-heated tank and the cold-water tank. The house's cold-water supply remains unaltered, while the hot-water cylinder is fed with solar-heated water from the solar tank.

This is a simple and economical way of organising solar heating and requires a minimum of alteration to the existing plumbing of the house, which is an important point. It is also suitable for use with various types of solar panel. The result is a link-up of the solar collecting arrangement shown in Figures 10 and 11 via the arrangement in Figure 13, added on to the household supply shown in Figure 15, the final result emerging as shown in Figure 16.

The indirect system. The indirect system works in a similar way (see Fig 14), the difference being that the solar-heated water does not pass directly into the hot-water cylinder but

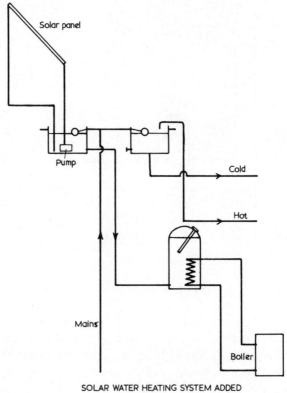

Solar panel

Pump

Cold

Hot

Mains

Boiler

SOLAR WATER HEATING SYSTEM ADDED

Fig 16

goes through a coiled heat-exchanger fitted inside the hot-water cylinder. This method is not so efficient as the direct method, due to loss of efficiency at the site of the coil; this loss is of the order of 15 per cent. Nevertheless, it allows antifreeze to be introduced to the circuit and is a system favoured by some firms.

Panel Direct to Cylinder

Those systems which do not include a solar-heated tank but lead the warm water direct from the solar panel to the hot-water cylinder also have a direct and indirect method.

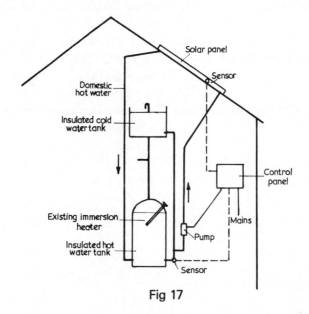

Fig 17

The direct method shown in Figure 17 uses the actual warm water direct from the solar panel run into the hot-water cylinder. The indirect method has a double coil in the hot-water cylinder (Fig 18), so that no actual mixing of the solar-heated water and the mains water takes place.

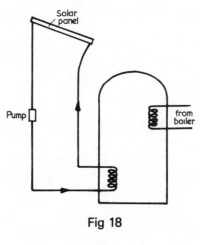

Fig 18

There are variations of the above system, some of which involve running a heated coil direct from the solar panel. With methods like these one must be very careful that heat is actually being taken to one's hot water system and not inadvertently led away from it! Some hot-water cylinders can suffer from having more holes bored into them than they are designed to accept; and the whole hot-water system can be capricious—the less it is altered the better.

3 Snags and How to Overcome Them

Many people install, or have contractors install, solar-heating systems and then find that their household plumbing has been upset. Maybe the shower will not work, or the water in the taps loses its power, or the central-heating system refuses to work normally. All this confuses and dismays any layman and may, alas, confuse the plumber, who should never have allowed it to happen in the first place.

Household plumbing is a simple concept and basically requires a sufficient head of water in the cold system to keep it functioning normally. Usually the bathroom shower, the highest and most sensitive piece of apparatus in the house, unless it has its own pump incorporated in it, will require a 4ft (120cm) head of water to enable it to operate effectively. This means that the outflow of the cold-water system in the attic must be at least 4ft above the level of the highest position at which you require a spray of water to come out of the shower.

Showers are particularly sensitive in this respect and also in their mixer taps. Mixer taps depend upon having similar pressures in both the hot and cold water inlets; so any unintentional reduction in the cold-water pressure means that the hot water in the mixer tap is at an advantage, forcing itself into the shower at the expense of the weaker cold-water pressure: consequently the shower sprays out scalding-hot water to the dismay and danger of the household.

Now although solar-heating systems do not usually involve altering the building's cold-water tank itself, the effective pressure can be reduced by various measures that may be adopted when installing solar heating. Having a secondary solar-heating tank at a lower level than the original cold-water tank may do it, as may withdrawing cold water with a join from a level lower than the tank, or narrowing the pipework below the tank. All these errors can easily creep into a system without one being aware of it. It is worth paying close attention to these points to avoid upsetting the normal workings of the household plumbing. This particularly applies to the siting of the solar-heating tank, which should be on a level with the existing cold-water tank no matter how far apart they may be.

Equipment Faults

Poor plumbing equipment should not be a worry, as apart from the solar-heating panel no equipment new to plumbing is involved. Thermostats, ballvalves, pumps, etc, have all been in use for a long time even though individual items occasionally fail. Even pump failure causes no disaster—it only means that the system quietly fails to operate. Pumps are guaranteed for at least a year and can easily be replaced. Thermostats seldom go wrong and any failure to function correctly is usually due to faulty setting. This aspect will be dealt with in the chapter on installation.

Poor functioning of the solar-heating equipment is, on the other hand, a common complaint. Some people find that their heating system only works in strong sunlight. This can be due to poor siting of the solar panel, using too small a panel for the size of the household, having a cumbersome panel containing a large volume of water which consequently takes a longer time to warm up, or the thermostat being set with its probes at too great a temperature apart, thus making the system unable to function in comparatively small temperature differences.

To take advantage of our capricious weather it is wise to have a solar panel which contains a small volume of water which can heat up quickly, and the thermostat sensors adjusted to allow a short burst of sunshine to be exploited. The adjustment of thermostats will be discussed in more detail in the chapter on installation.

Solar tanks will cool quickly unless they are adequately insulated. At least 4 inches of glassfibre should surround them.

Finally the pump should be checked to make sure that it is adequate for its job. It is particularly important to ensure that it can produce sufficient head of pressure. This will be stated on the pump's specification and must be greater than the height of the outflow from the solar panel above the water level in the solar-heating tank.

Make sure that an anti-syphon vent has been fitted to the system (see Chapter 5 on planning a system). Unless this is done the fall in temperature will cause the valve to open and instead of draining the panel it will attempt to drain the solar-heating tank. As this is constantly being refilled from the mains, via the ballcock, water will continually be syphoned out into the gutter — until the frost puts an end to the whole business.

Frost Damage

Frost damage is a particular hazard to solar panels. For as the temperatures drop on a frosty night, any water contained in a solar-heating panel could be turned to ice and expand, damaging the panel. At least the circulation through the panel would have stopped as the sensors would have told the motor there was no point in pumping water at the prevailing temperatures; but as well as this the panel must be effectively drained when the temperature falls anywhere near freezing. This is done by fixing a frost valve to the panel. This valve opens automatically as the temperature nears freezing point and allows the water to drain from the panel. (This is described in detail in Chapter 5, on planning a system.)

Workmanship Faults

Bad workmanship is obvious though not always entirely blame-worthy. Even in these days joints do leak and allowances have to be made for plumbers who are asked to do work to which they are not accustomed.

A break in the continuity of any roof is a potential source of trouble. Where a hole is made in the roof of a house, great attention must be paid to the flashing and sealing to ensure that the rain does not come in.

The incorrect installation of pumps is a source of error, and a ·dismayed householder can sometimes find water being pumped to the most peculiar places. Reference back to the original plan should correct this mistake.

More serious mistakes can occur when water cylinders develop bulges. This happens when an indirect heating system is used and a coil heat-exchanger is introduced into a cylinder which has not been designed to take it. Boring extra holes into cylinders which have not been designed for them weakens their structure and can cause distortion and bulging. This is a strong argument against using an indirect form of heating. (The simpler direct form also has the merit of needing fewer joints, which allows fewer things to go wrong.)

Planning Faults

Planning mistakes can easily be avoided as basic household plumbing is a simple concept and one is only adding an extra sideline to a part of it. Nevertheless, in some houses, especially older ones, it is not always easy to see which pipe goes where. In some older houses that have undergone modification mistakes can be made and you might even find pipes that literally go nowhere.

To avoid mistakes, the basic principles must be applied: site the new solar tank on the same level as the house's cold-water tank, make sure there is no diminution in the diameter of piping used, and see that no water pressure is 'stolen' by side outlets being sited where they should not be.

Failure to pay attention to these points can result in poor water-flow from the household taps, leading to seemingly unrelated complaints such as the washing-machine taking too long to fill or the dishwasher not cleaning the dishes properly.

Corrosion is a problem in some systems, as the joining together of two different metals, such as copper tubing to a galvanised tank, can set up an electric current between them and cause corrosion to set in. This problem has been largely overcome by using plastic components.

Systems which use the house's cold-water tank as a pre-heat solar tank should be avoided, partly because it would be wasteful to heat water destined for the cold-water taps and partly because warmed water in the cold-water taps is unpleasant. A further objection is that warm water in an open cistern allows the growth of bacteria and algae which can be a health hazard. A proper solar tank should be virtually closed to minimise these troubles—not completely closed, as access for adjustment of the pump, etc, is necessary. Algae do not grow on polypropylene though they do grow on polythene. Finally, if the cold-water cistern were used as a solar tank the ball valve would melt. Solar tanks require a special hot-water ball valve.

Building Regulations and Planning Permission

There are as yet no formal building regulations covering the installation of solar-heating equipment. It has been cautiously suggested that people should inform the local water authority when installing apparatus of this kind and the National Water Council can condemn any solar-heated water installation which does not conform to their standards. Their main concern is to eliminate health hazards and risk of contamination to the public supply.

Should any of the apparatus involve the use of components filled with antifreeze, the water authority should certainly be informed.

Cylinders with solder-seamed heat-exchangers are not acceptable to the National Water Council. If antifreeze solution is used you must check with the cylinder manufacturer that the coil in the cylinder has no joins whatever so there is no risk of leakage of antifreeze into the water. The fact that you seldom drink water from the hot taps is no reason for complacency.

At the time of writing no planning permission is required to install solar-heating equipment unless the house is covered by

37

an article or direction, or is a Listed Building. Planning permission is required if any part of the installation crosses the local 'building line' and this can apply to solar panels if they are fitted to the front of the building. Even if they do not cross the building line, they would involve a change in the appearance of the building and might be considered an eyesore.

Similarly, although no permission is needed to install solar panels on the ground in your garden, it would be necessary if you wanted to mount them on a tower or some other structure, possibly on the garage roof. It would be important to make sure that you were not keeping sunlight from your neighbour. Planning permission is also required if solar panels are used as part of the construction of the roof, or if the construction of the roof of the house is altered in any way. These matters can be discussed with the local planning officer, who will need to know what the panel looks like and how it is constructed. All the technical leaflets can be taken to him, including a rough sketch of how the panel or panels are going to be installed. It would also help if he could be told where he could see similar panels already in use.

Most uncomplicated installations will cause the planning officer no trouble, but where some point of concern is raised it may be necessary to make a formal planning application. This must be done on the correct planning application form, together with detailed scale-drawn plans and including modifications suggested by the planning officer.

The easiest way of achieving this is have your builder and his architect undertake it, but it is possible to do it yourself. In this case it is necessary to read the Building Regulations, 1977, and their subsequent amendments, before sending in the application. A decision on an application usually takes a minimum of two months. If your plans are turned down, it is always possible to modify them for resubmission: and if they are still rejected you have the right of appeal to the Department of the Environment.

You must also be prepared to have the building inspector come to visit the installation to make sure that all the regulations are being carried out correctly.

Building Society Permission

If you are buying your house through a building society, the society will need to be informed of any constructional alterations that are proposed. No difficulty should be experienced for routine installation of solar heating but all the same the society's permission should be obtained.

The building society's chief worry would be whether the solar-heating apparatus was going to weaken the structure of the house, or alter its appearance in such a way as to decrease its sale value. With this in mind, societies prefer commercial panels to the home-made variety. An adequately installed solar-heating system used to back up a conventional heating system would tend to increase the value of the property.

To arrive at a decision the building society need the detailed plans to show exactly what has been done to the property. This should include photographs of the solar panels, obtained either from the manufacturers or from properties with similar installations. They will need to be reassured that nothing is being done to render the property unconventional.

4 Cost

The true cost and value of any piece of long-term equipment is almost impossible to calculate in an age of fast inflation. Claims made by solar-heating firms describe savings of 40-60 per cent in conventional fuel bills and declare that the apparatus will pay for itself in five to fifteen years. Parts of these claims are based on calculations assuming that the price of conventional fuel will rise by at least 10 per cent per annum and to translate such predictions into hard figures is next to impossible.

Savings to be Expected

Each year 1,000 kW hours of sunshine fall on each square metre of land, so in theory this amount of energy can be picked up by each square metre of solar panel. This assumes that the panel is sited in the best possible position. We have seen earlier that the highest efficiency of solar panels is about 40 per cent, in which case the very best we can gain if the system is running at its most economical (which it never is), is 400 kW hours energy. If this saved the equivalent amount of electricity at, say, 2½p per unit, the saving would be £10.

This means that each square metre of solar panel could save £10-worth of electricity every year. The average 4 square metre solar panel would save £40-worth of electricity per year.

An average individual may use 4,000 kW hours of energy to heat his water each year. With 1 kW hour of energy provided by 1 unit of electricity at 2½p per unit, his water-heating bill would be £25.

This kind of sum appears to be the maximum that can be saved per individual per year by solar heating, assuming again that it was a saving in electricity. If the same amount of heat was obtained from gas or oil or cheaper fuel, correspondingly less would be saved by solar heating.

One point to remember is that central-heating boilers are very inefficient when used below their rated output; this occurs in summer when the radiators are turned off and the boiler thermostat is turned to minimum as only a small amount of hot water is required. In these circumstances it is often more economical to switch off the boiler and use the electric immersion heater. This is one reason why savings from solar heating are based on electricity prices regardless of the fuel used.

It has been explained that for maximum solar-panel heat gain, water should be drawn off constantly: in a normal household this is never the case, so the overall saving would be reduced. When looked at in this way, and especially when compared with savings you could gain (on paper at any rate) by putting money in a building society, the 'savings' claimed can be made to mean anything.

One figure was quoted for a solar-heating system costing £650. The equivalent sum of money kept in a building society at 6½ per cent per annum would reach £1,000 after twenty years which would be the break-even point if the money were spent on solar heating. If fuel costs rise 10 per cent per year, break-even point would be reached in seventeen years. Solar-heating advertisements can be misleading and their claims must be carefully evaluated. However, with inflation running at about 20 per cent, the buying power of that money in the building society diminishes constantly and is not made up by the interest you receive. When you compare this with the very real saving made when a modestly installed solar heating system cuts your fuel bill by nearly half, the whole thing begins to make sense. Savings are even more if circumstances allow the fitting of a solar tap, when for most of the day over eight months of the year no other form of water heating is required at all.

Thus investment in a solar panel should be producing a continuous dividend while still retaining the original equipment. There are signs that the Government are slowly becoming aware of the need to save energy and to encourage solar heating. There is no VAT on solar-heating equipment and as already mentioned there may well be some tax rebate, up to half the original capital cost, if suggested legislation is passed by Parliament.

On balance it does appear that fuel consumption is reduced by at least one-third, and if fuel prices rise by 15 per cent per annum the capital cost will be recovered in five years, if an inexpensive system is used.

The Cost of Installation

The price of individual systems can vary tremendously. The cheapest installations, at the time of writing around £30, involve the use of secondhand or scrap equipment installed by yourself. Old radiators painted black have been used as collecting panels, though their bulk and weight make them difficult to install on the roof. At the other end of the scale, refined commercial installations can cost about £1,500. It is said that contractors can install this type of equipment in about two days and that the money can easily be borrowed from one's friendly bank manager: it seems to me that the bank manager would have to be unusually amicable to enter into an arrangement to finance equipment which is still partly experimental.

A more reasonable approach, the one most likely to be economical and efficient, seems to be to allow for a simpler form of apparatus costing only a few hundred pounds and to have it installed by the local plumber.

As a rough guide the prices for individual items in 1978 were as given below. Cost will of course change frequently but these give some idea of the relative size of the various areas of expenditure.

Solar panels, £27-£32 per sq metre (the larger the panel the cheaper per sq metre). Solar panels without their frames are approximately one-third cheaper.

Differential temperature controller, £38-£50.

Pump, £26-£36.

Frost valve, £5.

Angle brackets for fastening collectors to roof, £5.

Hose, flexible or rigid, £1 per metre.

Compression fittings, £5.

Anti-syphon kit, £4.

Lead flashing 150mm wide, £2 per metre.

As a package deal the average family house might require equipment costing between £200 and £250 (1978 prices) for a simple system not involving a heat exchanger or sealed systems.

Maintenance is minimal as there is very little to go wrong. The main moving part is the pump, which should last for ten to fifteen years; it should carry a guarantee and is easy to replace if necessary. The pump uses very little electricity and usually consumes the power equivalent of a 60-watt light bulb, though the cost of this is further reduced as it only runs when there is profitable heat to be gained.

5 Planning a System

Planning your own solar-heating system depends very much upon your inclination and makeup. Armed with the information from the previous chapters, you can weigh up the merits of the various systems advised by the brochures and representatives' visits. The active DIY man may well prefer to knock up his own unit from scrap radiators and other material for £30, while the more affluent may succumb to the temptations in the brochures and opt for a professionally installed system costing from £1,500 upwards.

In practice you are likely to discard the cheapest thermosyphon system as inefficient, and choose the more usual pumped system. Whatever you finally select it must be sited to obtain the best possible results and be of adequate size to produce a reasonable amount of water for the household.

It is not, of course, necessary to install a complete solar system for the whole house. A smaller installation to supply hot water to a solar tap in the kitchen or outhouse might be all that is required, in which case a very cheap and simple system will be good enough.

Hot Water Requirements

All households vary in their hot water requirements and accurate sizing is impossible. As has been mentioned solar heaters work best when water is continually withdrawn throughout the day and this just does not happen with most families. Any attempts at giving specific sizes you will need, are therefore only approximate.

As a rule of thumb, one square metre of solar panel will be required per adult, two children counting as one adult, while the solar tank capacity should allow 45-67 litres (10-15 gallons) of water per adult per day. A quick calculation will show this to mean most families require a solar panel of 2-3 square metres in area with a water tank of about 225 litres (50 gallons) capacity. This new water tank, separate from and in addition to the ordinary household water tank, comes in one piece and is mounted in the roof space, usually near to the solar panel. The panel itself can come as one large unit or several smaller units joined by pipes, depending upon the space available for it.

Which Type of Equipment?

This is the most difficult decision that any householder has to make. The fact that so many types of systems and equipment are available means that no one type is perfect. Provided the equipment is reliably manufactured you do not necessarily obtain the best deal by paying the most money, and remember there is still some controversy about the financial advantage of installing solar heating. This is mainly based upon the high cost of the more elaborate systems. Heat is a capricious commodity and can be lost if an over-complicated installation is employed. All the indirect methods suffer from loss of efficiency during heat transfer from the solar-heated water to the normal water supply of the house, so that at the present time the cheaper and simpler direct method appears to be the most satisfactory.

If the total cost can be kept low with a simple and fairly efficient set of apparatus, then the individual parts can always be replaced if a more effective component comes onto the market.

General Considerations

Aim for simplicity. The simpler the system installed, the cheaper it is, the quicker it can be done, the less there is to go wrong and the less upset it causes to the household in general.

In all construction work expense and weight go hand in

hand and it is certainly an advantage to have a fairly light solar panel on the roof. It is very much easier to manhandle into position and there are fewer things that can go wrong with it.

Throughout the whole operation make sure there is no diminution in the diameter of the pipe leading from the solar tank down to the hot-water tank, and make sure that no other pipes are allowed to run off this connection. Failure to observe these rules may mean upsetting the existing pressures in the water system and can upset the flow to showers and taps, making the taps sluggish and preventing the showers from working properly.

Solar Water Tap

Is a solar water tap to be fitted? This will deliver water from the solar-heated tank down to either the bathroom or the kitchen, or both, enabling the household to use solar-heated water direct. If the distance from the tank to the point of use is short it is a very worthwhile installation, as the saving in fuel can be tremendous and, as mentioned before, continuous use of this water can improve the efficiency of the system. In many houses, alas, the distance is too great for this to be economic.

The Solar Panel

Having decided upon the type of solar panel preferred, and calculated its size, plan its position.

The best place to put a solar panel is on the roof. It can be laid over the existing tiles, or on a proposed new building. Providing allowance is made for expansion it can be used as part of the roofing itself, with considerable saving in the cost of the tiles. It works most efficiently if the roof is sloping at an angle between 45° and 60° from the horizontal. It can be laid flat on a flat roof, but this way its efficiency will be low except in high summer.

Solar panels work more efficiently if laid on an adequate bed of insulation, as up to 30 per cent of the heat can be lost in this manner. A thick layer of fibreglass enclosed in heavy-duty polythene makes an adequate insulating layer for this purpose.

Remember that if the panel is left without water in it during the summer it becomes too hot to touch, and the temperature can easily melt polystyrene insulation. Polyurethane foam (from Coolag Ltd) has a higher melting point. Fibreglass is satisfactory as long as it does not become saturated with rain-water or condensation, which is why it should be encased in heavy-duty polythene.

Siting the panel may present some problems. Obviously a south-facing roof is preferable but in fact satisfactory results can be obtained on a roof facing anywhere from south-east to south-west. If a suitable roof is not available the panels can be laid vertically over the front of the house. In this case twice the area will be needed to make up for the loss of efficiency and one may need to obtain planning permission, especially if the installation comes up or goes over the local building line, or alters the appearance of the building.

It is important that the site should be free from overhanging trees, or indeed anything that would obstruct the sunlight for a considerable part of the day. It must also be free from any likelihood of falling debris. If no suitable site is immediately apparent, a little ingenuity can often devise one. Successful solar installations have been mounted over canopies, perhaps over a porch, or mounted on a frame over the garage or even in the garden. Frames must be strongly constructed and offer sufficient protection from wind and storm. Remember that the wind will try to get round behind them and the relatively large areas involved in solar collection make them specially vulnerable to wind damage.

It is an advantage to keep the whole system compact to lessen the need for expensive pipe work. Also you should be able to reach the solar panel easily to check the frost valve (see page 53) from time to time.

The Solar Heating Tank

Assuming that the solar-heating panel can be satisfactorily mounted, check that the solar-heating tank (for details see next chapter) will actually fit into the roof space. Is there room to site it fairly near the solar panel as well as keeping it at the same level as the ordinary cold-water tank? This may mean building a special platform strong enough to support it. Details of how to construct one will be given in the next chapter.

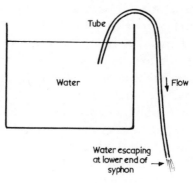

Fig 19

Anti-Syphon Vents

At this point consideration must be given to anti-syphon vents. The syphon principle was known to the ancient Egyptians. What happens is this. If one end of a tube filled with water is put in a tank of water, while the other end is outside the tank and held at a lower level, water will flow down the tube and drain the tank. It only stops when the level of water in the tank has become so low that air is admitted to the upper end of the tube.

This principle is simple and has been taught to generations of schoolchildren. It has been adapted for innumerable industrial uses, and it is a principle that can creep into plumbing almost without one being aware of it.

Its importance in solar heating is that unless you are very careful you can unwittingly create a syphon. It can happen when, for example, a solar panel's frost valve opens below the level of the water in the solar-heated tank: as fast as the frost valve tries to empty the solar panel, it is refilled with water drawn from the solar-heated tank, which of course is constantly being replenished by water from the mains. This situation can go on forever — until either the temperature rises and the frost valve closes itself, or the frost becomes so severe that the water in the panel freezes and damages it. Obviously it is a complication that can only occur in solar panels sited partly or wholly below the level of the water in the solar-heated tank.

The problem is overcome by fitting what are called anti-syphon vents. Under certain circumstances these allow air to be drawn into the system instead of water, thus stopping the syphon effect.

The anti-syphon tubes may be lengthy and it is important to allow sufficient space for these to be fitted if necessary. And they are essential if the solar panel is fitted below (partly or wholly) the level of the solar tank, when one or more anti-syphon tubes must be installed.

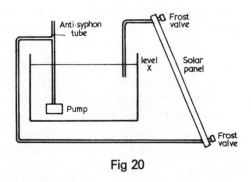

Fig 20

A solar panel mounted above the solar tank creates no syphon problems so no anti-syphon system need be considered.

In Figure 20 the solar panel can be seen partly above and partly below the solar tank. The anti-syphon tube is fitted to the pump side of the solar panel's circulation, so that when the system drains in frosty weather via the lower frost valve, air will be allowed into the lower water channel, preventing it syphoning out of the solar tank.

Notice that the upper water channel is run under the level of the water in the solar tank. If this were not done, air would be drawn into this channel every time the pump stopped and the water level in the solar panel would fall to level X — the same level as the water in the solar tank. This is because water will always find its own level.

Syphoning out of water in this channel is prevented by the upper frost valve, which lets air into the panel when the temperature falls near freezing, thus preventing a syphon forming in this tube.

To repeat: when frost valves are adjusted it is preferable to have this upper valve opening at a slightly higher temperature than the lower frost valve, to prevent the formation of a temporary syphon.

The anti-syphon tube fitted on the pump side of the solar-panel circulation must be long enough to exceed the height to which the pump is designed to pump water. This can be found by looking at the pump's specification.

If the anti-syphon tube is not made long enough, then some of the water destined for the solar panel will be pumped up to this tube and will cause a disastrous flood. There may be sufficient height in the attic or roof-space to allow for this head of pressure, and the anti-syphon tube can be angled to make use of all the available height. If the height available is insufficient, then the anti-syphon tube can project through the roof and carry on as high as necessary.

It is possible to reduce the height to which the pump will pump water by fitting a choke on the pump; this is in effect a constriction on the pipe to lessen the pump's effort, though it must not be lessened sufficiently to interfere with pumping the water through the solar panel.

Where the solar panel is installed completely below the level of the solar tank two anti-syphon vents will be necessary — see Figure 21. The first should be on the pump side as described already, and the second on the pipe bringing water from the solar panel back to the tank.

The necessity for these anti-syphon vents is not always apparent to the plumber, yet it is vital that they are installed

50

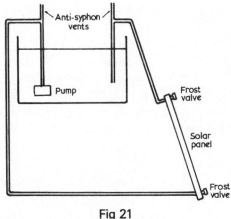

Fig 21

correctly. It is sometimes necessary to insist to a plumber that he installs them although he may swear that they are unnecessary!

Overflow Pipe

At this time an overflow pipe must be planned leading from the solar tank to a point visible from the outside of the house.

The Pump

All efficient solar-heating systems require a pump. Only a simple pump is required but the submersible type has advantages. One is that the heat generated is not wasted but warms the water; another is that it needs no fixing but just lies in the bottom of the water tank. Such pumps are commonly used to operate fountains in fish ponds.

All pumps should come with leaflets showing details of their performance. You need to know their maximum flow and maximum head, the voltage required and the electricity consumption.

Maximum flow is quite simply the amount of water that the pump will pump in a given time. It should be able to pump the

51

entire contents of the solar-heated tank through the panel every ten minutes, and to pump out the contents of the solar panel every minute. It is important to check that the pump is adequate in this respect; too feeble a pump will cause a fall in overall efficiency. In practice a pump with a flow of 0.6 litres per second (8 gallons per minute) will be adequate.

The maximum head is the highest point to which the pump can raise water and is usually in the region of 2-3 metres (6½-10 ft). Whatever the maximum head, it must be more than the distance from the surface of the water in the solar tank to the outflow from the top of the solar panel. Sometimes, with a large solar panel, this distance may exceed the capacity of the pump and the solar panel must be fitted lower down the roof accordingly, or a more powerful pump obtained.

The pump's voltage should be suitable for the normal mains electricity supply, usually 200-250 volts at 50 cycles alternating current. Some pumps operate at 12 volts and are powered from the mains in the usual way though through a transformer. The transformer can be placed next to the controller.

These pumps are usually equipped with an automatic cutout which operates when the surrounding tank water is approximately 50°C (120°F). This is a safety device to prevent over-heating of the water and will only operate during very hot conditions when no water is drawn from the storage tank.

Electricity consumption by your pump will be minimal, in the 60-90 watts range, equivalent to a low-powered light bulb. With a submerged pump, part of this energy is recovered as heat—the running of the pump heats the surrounding water.

The Controller

As seen in Chapter 2, it is the controller that is the brains of the whole solar-heating circuit. Controllers have been specially devised for this task; they come with various refinements, but they all do the same basic job. Nevertheless, a controller must be selected with great care as the whole system will depend upon its reliability for many years.

The basic controller is housed in a small box which has two

temperature-sensing probes attached to it by long insulated wires. One probe is fitted to the water outlet from the top of the solar panel and the second is fitted into the solar tank, about a quarter of the way up from the bottom and fairly near to the submerged pump. A movable knob projects from the front of the controller and a pointer on this control knob usually indicates a calibrated scale.

What happens is that the sensors relay the temperature to the controller, and its knob can be adjusted to make the pump operate when there is a certain difference between the two temperatures. Thus, if the knob is turned fully anti-clockwise, the pump will operate when there is a very small difference in temperature between its sensors; while at the extreme clockwise end of the scale, the pump will only operate when there is a very large temperature difference, usually 8°C.

The actual adjustment of the controller will be dealt with in Chapter 6, but as long as the machine is well made it is not necessary to have a calibrated dial upon it—once set it seldoms needs changing.

The electricity is led from the mains through the controller to the pump, and an indicator light on the controller shows when the pump is operating.

When there is useful heat to be obtained from the sun the temperature differential will switch on the water to start the apparatus working. These sensors are fundamental to the successful working of the scheme. They stop the motor instantly if the sun passes behind a cloud, thus preventing heat waste and also saving the electricity used to run the motor.

Frost Protection

With all installations other than the indirect type filled with antifreeze, some form of frost protection is vital. As the temperature falls below freezing, ice will form in the solar panel and expand, quite possibly cracking the panel. This must be prevented. The simplest method is to use a frost valve which will empty the panel as soon as the temperature approaches danger point.

This is an important and ingenious mechanism, a plastic tube which incorporates a valve controlled by a thermostat. The narrower end of the valve can be plugged into the solar panel and it incorporates a side arm which will either admit air or let out water according to its location, allowing the solar panel to drain.

The wider end of the valve incorporates the regulator, which can be turned by hand to adjust the temperature at which the valve will open. The regulator is calibrated, usually from 0 to 10. When the frost valve is connected to the solar panel and the apparatus is working, the regulator is set fully clockwise at 0. When in winter the temperature falls to near freezing-point, the frost valve will open. This will allow the water to drain out of the solar panel before it freezes.

If the solar panel is mounted above its water tank, then the frost valve is mounted at the top of the panel; when it opens it admits air, allowing the water to drain back from the solar panel into the solar tank.

If the panel is mounted below the solar tank, the frost valve is sited at the bottom of the panel allowing the water to drain from the panel into the gutter of the house. The anti-syphon vents already mentioned prevent further water from coming into the solar panel.

If the panel is mounted only partly below the solar tank, then two frost valves are needed. The one at the top of the panel, which should be adjusted to open first, admits air into the upper part of the panel; the lower valve allows the water to drain out, helped by the anti-syphon tubes already mentioned. The valves automatically close themselves when the danger of frost is passed. The pump has already ceased working at this time as its sensors have told it there is no heat to be gained from the panel.

Frost valves seldom give trouble, though it is wise to check them before the onset of each winter. The actual closure is made by a rubber valve poppet, which under solar-heating circumstances seldom becomes worn as it does not have to open and close at frequent intervals. You can unscrew the valve and remove the tensor rod, which allows the valve to open. The opening and closure of the valve can be checked by pushing the pin. At the same time debris can be removed by flushing with air or water. When it is clear you can check that the valve is sealing properly by blowing air into it with the pin pushed in. If, through wear, the valve will not seal, the valve poppet can be replaced, but for this special tools are required. If your plumber cannot help, the valve may need to be sent back to the manufacturers.

Occasionally you are made aware that a frost valve has become blocked by debris and is unable to close properly by the fact that it starts to leak, in a steady drip or even a stream of water. You see the leak or notice water coming down the appropriate downspout at a time when there is no rain! You may also be conscious of the noise of running water constantly replenishing the level in the solar tank. Alerted by these signs, it is a simple matter to switch off the pump, remove the frost valve and unscrew it to give it a thorough clean. The debris is easily dislodged and the valve reassembled and put back in place.

Pipework

The pipework should be kept to a minimum to save losing heat, to keep down costs and to keep the installation neat. Flexible industrial-grade clear PVC hose offers the simplest possible method of installation. This piping fits directly on to the solar panel and as the connections are made inside the roof, the procedure is simple.

Should the panels be wall-mounted, the piping should be lagged, partly to conserve heat and partly to shield it from ultraviolet radiation. Rigid Class E PVC pipe with solvent-type fittings may offer a neater installation and it last longer than flexible hose.

For thermosyphon circulation the bottom of the solar water tank should be at least a few feet above the solar panel top. The minimum pipe diameter is 22mm (7/8in) or 28mm (1¼in) for long runs. In a pump system 15mm (¾in) pipe is adequate.

Finally, always bear in mind three general points when planning a solar-heating system.

Corrosion

Remember that if two different metals are used in a plumbing installation where they are in contact with water, an electrical circuit is set up between them, causing the less inactive metal to corrode. Thus a copper washer should not be used on an iron cylinder. Always keep this principle in mind when planning any project. Fortunately the problem is overcome when plastic is used, as it is electrically inert.

Ballvalves

Some ballvalves do not stand up to hot water and it is important to make sure that a heat-resistant ballvalve is used whenever there is any possibility of it coming into contact with water that is anything other than cold. Solar-heating systems can over-heat, particularly in the tropics, and part of the solar panel will need to be shaded. As mentioned later, different systems can be used in tropical climes, and local advice must be sought in these situations, though the basic principles remains the same.

Algae

Wherever there is water, algae tend to grow. These minute green slimy filaments of vegetation reproduce even under the most adverse conditions. Fortunately they do not grow on polypropylene—but they do grow on polythene. The slime can easily be wiped away and the growth is reduced by covering the tank well to exclude light. Nevertheless, the warm conditions in a solar-heating tank do favour the growth of algae and this is one argument for avoiding a totally enclosed tank, as the inside would be virtually impossible to clean.

6 Installation

There is nothing complicated about installing a solar-heating system, though for some parts a basic knowledge of plumbing is neeeded.

The basic compression joint which is detailed below is simple. Nevertheless, the plumbing in some houses is far from simple and the penalty for a small mistake can be a disastrous flood. Re-routing pipes to the hot-water cylinder can be a daunting procedure for the uninitiated. Work must proceed methodically and it is possible to do a great deal of the preparation yourself, only calling in a plumber to connect up the actual pipes.

The Solar Heating Tank

This new tank will normally be put in the loft. Most tanks up to 225 litres (50 gallons) capacity will pass through the normal 61 × 61cm (2 × 2ft) loft opening and a tank made of polypropylene can be flexed slightly by means of a rope to ease it through the opening, should it be a tight fit. The rope must then be released immediately. The tank must be sited carefully, if possible on a level with the existing cold-water tank and fairly near to the solar-heating panel. It must stand on a firm level surface such as a piece of chipboard 2cm (¾in) thick and 5cm (2in) larger all round than the tank. This should be placed across the joists, preferably over an internal load-bearing wall.

Often it will need to be put on a raised platform and in view of the weight of the tank three-quarters full of water, carefully and solidly made. This is a standard construction (see Fig 22). Two pieces of timber 5 × 10cm (2 × 4in), slightly longer than

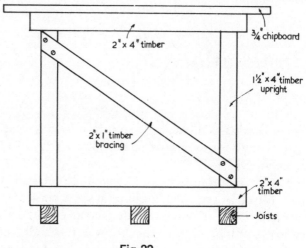

Fig 22

the length of the tank, are laid side by side, flat side down, slightly less than the width of the tank apart, over at least three joists; again, one of the joists should preferably be part of an internal load-bearing wall. Four upright pieces of timber, 3.25 × 10cm (1½ × 4in), are stood on these timbers to form the main supports of the tank. On the upper ends of these are placed two more pieces of timber, 5 × 10cm (2 × 4in), matching the first two that were laid over the joists. Each side must be braced with a diagonal piece of timber 5 × 2.5cm (2 × 1in), as in the diagram, and the platform completed by fastening to the top a piece of chipboard at least 2cm (¾in) thick and 5cm (2in) all round larger than the base of the solar-heating tank.

Correctly located water connections must now be fitted. If you have doubts about your competence as a plumber, this part of the installation should be undertaken by a competent tradesman.

Holes drilled in the plastic tank must be carefully marked in pencil or crayon and not scratched on. Care must be taken that all pipework joins the tank at right angles and that there are no internal lugs or strengthening bars where the pipes are to be fitted. Drilling is done with a hole saw in a power drill, and it is essential to support the tank wall during drilling. Clean holes must be made, all scraps of plastic carefully removed and the holes cleaned with steel wool or emery paper.

In modern plumbing, compression joints are used. These are much simpler than the old soldered joints, as no heating is required. To make a joint by the compression method (Fig 23), the nut and compression ring are slipped on to the pipe with a little jointing paste or PTFE applied to the ring.

The end of the pipe is then pushed through the hole so that the ring fits firmly against it. Finally the nut is screwed on by hand, and after checking the alignment the join is tightened with a spanner or wrench.

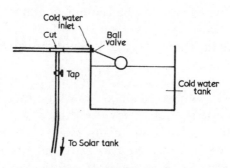

Fig 23

Three holes are needed in a solar-heated tank. The mains inlet should be situated 3.25cm (1½in) from the top of the tank and a high-pressure ballvalve must be used with a float suitable for hot water. The overflow pipe is placed 5cm (2in) from the top of the tank, while the third hole for the hot-water outlet is situated 5cm (2in) from the bottom of the tank.

At this stage an extra outlet for a solar tap can be fitted. Usually sited 5cm (2in) from the bottom of the tank, it is led down to the bathroom or the kitchen as desired and fixed to an extra tap in the bath, sink or basin. This task is usually undertaken by a plumber.

Pipework

As mentioned in Chapter 5, flexible industrial-grade clear PVC hose offers the simplest method of connecting up the various pieces of apparatus. It can be connected directly to the pump and to the elbow adaptors at the inlet and outlet of the solar panel. For solar panels mounted on the roof the connections are even simpler, as they are made inside the roof.

When wall-mounted solar panels are used, flexible hose can still be employed; but as it will need to pass outside the house, it should be lagged to shield it from ultraviolet radiation as well as to prevent heat loss.

Rigid PVC pipework is a neater method of connection, especially where pipe runs are in the open, as when wall-mounted solar panels are used. It also lasts longer than flexible hose.

The two types of pipe can be combined. When fixing rigid PVC pipe to walls, etc, make sure it is free to expand and contract along its length and that it is not constricted between two points.

The joints for rigid and PVC pipes are not difficult, and involve cutting the pipes to length, making sure the ends are square, removing the rough edges with steel wool or emery, and using the appropriate compression fittings. Solvent cement is brushed on to the two ends and the fittings are pushed on to the pipe. The compression fitting is then screwed up by hand and tightened with a wrench.

Rigid pipes are used for the anti-syphon vents as they are easier to fasten to the correct height.

The Cold-Water Inlet

The solar-heated tank may be fed either from the mains or from the cold-water system, depending upon which plan has been adopted.

If fed from the mains, proceed as follows. First, turn off the mains cold-water supply to the house. This is most important. It isolates the home from the mains and will prevent a flood from occurring should anything go wrong! Allow the water level in the main pipe to drop by turning on the kitchen tap and depressing the ball in the cold-water storage tank for a few seconds. This drains the water from the section of pipe on which you are going to work.

At a convenient point remove 1.25cm (½in) of this pipe to allow a T-fitting to be inserted, using a compression fitting for the joints (see Figs 24 and 25). A stopcock can be fitted between the T-connection and the solar tank so that it can be isolated if necessary. The far end of this pipe is led to the solar-heated tank, and connected to it and the high-pressure ballvalve with a float suitable for hot water. The connection is made by means of a swivel tap connector. All ballvalves with a large float have a metal backing plate and this should be fitted on the outside of the solar-heated tank, pointing down, to resist the force of water pushing up on the ball. If a Torbeck valve with a small float is used, this plate is not necessary.

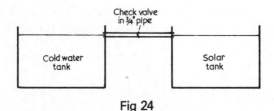

Fig 24

Some systems allow solar-heated tanks to be fed from the cold-water tank where it is at a slightly higher level. In this case a low-pressure ballvalve should be used, and if there is only a small difference between the level of the two tanks it will be necessary to use a larger diameter pipe, 22mm (¾in) and a 2cm (¾in) ballvalve to ensure an adequate feed rate.

If the original cold-water tank is so sited that the head of pressure is only just adequate to allow the household system to function, take care not to make matters worse.

If both tanks are mounted at the same level, then the solar tank can be fed without a ballvalve and a 2cm (¾in) connecting pipe can be used. The redundant hot-water feed from the

cold-water tank can be used to feed the solar tank if convenient
—though using this method it would be possible for warm
water to flow back into the cold-water system when cold water
is drawn off. It is, therefore, advisable to fit a non-return or
check valve in the pipe (Fig 24).

The Hot-Water Feed

We know that the household's hot-water cylinder is normally
fed by its own pipe from the cold-water cistern. In the solar-
heating arrangement it is fed with the warmed water from the
solar-heating tank.

Normally two pipes run from the cold-water tank, one going
to the hot-water cylinder and the other serving the cold taps
and services. To find out which is which, run first the hot and
then the cold water into the bath, and feel down inside the
cold-water tank to see which outlet is being used. When you
are quite certain which is the hot-water feed, drain the cold-
water tank. This is done by turning off the mains or tying up
the float valve to stop the inflow of water, and running off the
water via the cold bath-tap until it is below the outlets.

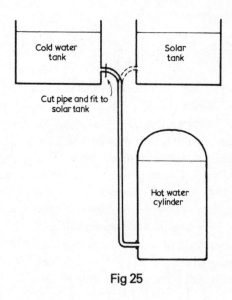

Fig 25

Lower the water level in the hot-water feed by running the hot bath-tap for a few seconds. The hot-water feed pipe can then be cut a few inches from the cold-water tank; fit a blanking plug to the end protruding from the tank (unless it is going to be used to feed the cold water to the solar tank). The other end of the feed pipe can be connected to the solar tank with 22mm (¾in) pipe with a 22mm (¾in) tank connector fitted to the hole that has been made 5cm (2in) from the base of the solar tank (Fig 25).

Sometimes these manoeuvres result in air-locks in the hot-water system. These are easily cleared by connecting a length of flexible hose to the hot-water vent pipe from the hot-water cylinder, and connecting the other end to the submersible pump which can be placed in the cold-water cistern. If the pump is run for a few seconds the offending bubbles will be seen emerging from the hot-water outlet in the solar tank.

Overflow Pipe

The solar-heated tank must be fitted with its own 2cm (¾in) overflow pipe to the outside of the building in a visible position. The pipe must, of course, run steadily downhill from its fitting 7cm (2½in) from the top of the solar-heating tank.

Mounting the Solar Panel

Solar panels are connected to tiled roofs by means of angle brackets which slide between the tiles. The outer ends are secured to the solar panel while the inner ends are fastened to battens fixed across the roofing rafters. The pipes from the solar panel are brought through the roof by removing two or more tiles near the inlet and outlet pipes and replacing them with lead flashing cut the same shape as the tiles and drilled to take the pipes.

It is vitally important that this work is done correctly. Tiling roofs to prevent leaks is a highly skilled business and it is dangerous for an unskilled person to work on a sloping roof. Solar panels are unwieldy and although comparatively light at

63

ground level, they can be difficult to manoeuvre into position. As they are easily blown about in a breeze, work must be undertaken on a calm day. At least two people with two ladders are required for this job.

Safety is all-important and the ladders should be tied securely to chimneys or windowframes. The ladders should be 1.75m (5ft) apart (a roofing ladder must be used to climb on to the roof).

Before mounting the solar panel, remember that any insulation must be fixed securely to its back.

The solar panel has a hole at each corner for the connection of pipes and frost valve. The actual joins are made by reducing bushes, which expand as the appropriate filling is pushed in place. Before going up the ladder push a reducing bush halfway into the opening at each corner of the solar panel.

The solar panel is placed in position on the roof and secured temporarily—with a rope round a chimney or some other convenient support, or thrown over the ridge and secured to the other side of the house. Strips of metal should then be pushed between the tiles just below each of the solar panel's four fixing points, while an assistant goes into the attic to see where they enter the roof space. To see them he may have to cut through any roofing insulation that is in the way.

It may be that they come up against a rafter, in which case the solar panel will have to be moved a little to one side. At this point you should make sure that the solar panel inlet and outlet pipes are also clear of the rafters. One angle-iron is then pushed out alongside each of the metal strips.

Next lift up each angle iron in turn and insert a 5mm (¼in) bolt from underneath each one. This is fixed with a lock and nut straight onto the angle iron, plus enough nuts and spaces to ensure that the tiles will not be lifted when the solar panel is fitted.

When this has been done the solar panel can be lifted on to the four bolts. A nut is put on each bolt loosely, but not tightened yet. When all four angle-irons have been satisfactorily attached and the solar panel is lying neatly on the roof, the inside portions of the angle irons are loosely secured to pieces of wood fastened across the rafters at right angles.

These fastenings are left loose until the inlet and outlet pipes are in position. These pipes are preferably at opposite corners of the panel, though if rafters or other difficulties prevent this, they may be placed on the same side.

Having decided where the inlet and outlet pipes will be, remove two or three tiles at each position to enable the elbow pipes to pass into the loft. These tiles are replaced with similar-sized 'tiles' of lead, on which are marked the position of the elbow pipe. Holes must be cut in the lead to allow the elbow pipe through. These holes are made slightly smaller than the pipe and enlarged with a tapered mandrel pushed up from below to form a lip.

The lip should be on the top side so that when the pipes are in position the lip can be tapped round the pipe to form a watertight seal (Fig 26).

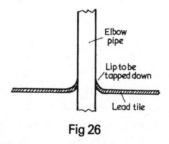

Fig 26

As an alternative, mastic can be used. It is essential that this is done carefully and correctly as there is nothing worse than a leaky roof to show for one's efforts.

Before all this is tightened up, the long end of the elbow pipe is pushed in through the lead tiles and manipulated so that the short end will go on the solar panel. A watertight seal between the solar panel and the elbow pipe is made by means of the rubber bush which expands as the elbow pipe is pushed in. To do this, position the rubber bush so that it has 0.5cm (¼in) to go before it is fully home, and then insert the short end of the elbow pipe, pushing it right through the rubber bush. This expands the bush, making a watertight seal, and also takes the bush right in up to the shoulder.

The fastenings are now made secure and the remaining two solar panel pipes are sealed off, one with a blanked-off pipe and the other with the frost valve. If the solar-heated tank is below the solar panel, the frost valve goes at the top and the blank end at the bottom. If the collector is below the solar-heated tank, the frost valve is fitted at the bottom and the blank end at the top.

If the solar panel is partly above and partly below the solar-heated tank, two frost valves are fitted, one at the top and one at the bottom of the solar panel. As described earlier, the upper frost valve should be set at one (1) on the regulator and the lower frost valve at zero (0) to prevent a short-term syphon effect. Anti-syphon vents must be used with the latter arrangements.

The wire from the sensor which is attached to the outlet elbow should be slipped between the lead tiles into the roof space and pulled through.

Solar Panels Mounted on a Framework

If a solar panel is to be mounted on a framework, this should be rigid and self-supporting and never rely on the solar panel for stiffness. Suitable frames can be made using slotted angle-iron, 3.25 × 3.25cm (1½ × 1½in), preferably galvanised but otherwise well painted, with nuts, bolts and washers of a similar material. This framework should be firmly attached to its surroundings with Rawlplugs or other proprietary fastenings. These fastenings must be secure. Solar panels in this position must be laid on a bed of insulation and backed by a firm protective structure preferably made of wood.

Any exposed pipework must be securely fixed to the wall, insulated, and protected from sunlight.

Connecting the Solar Panel to the Solar Water Tank

Flexible PVC hose, or rigid connections if preferred, are connected to the solar panel's outflow pipe and secured with a jubilee clip. The other end is directed to the solar-heated tank, into which it should project 2.5-5cm (1-2in), being cut off just above the water level.

The pipe will expand and contract according to the temperature of the water within and allowance should be made for this. It must be allowed to contract and expand along its length and must not be constricted between two points.

If the panel is partly above and partly below the solar tank, this pipe should project 2.5-5cm (1-2in) below water level.

Pumping Water to the Panel

The arrangements for this depend upon whether the solar panel is above or below the tank and whether the pump is submerged in the tank or is mounted free.

Solar Panels above the Storage Tank. If a submerged pump is used, flexible piping is connected to the pump outlet (the pump sitting on the bottom of the tank) and is led up to the elbow joint at the lower end of the solar panel (Fig 27). The pipe should preferably not run downhill at all as it goes to the solar panel, or there will be a loop of water left in when the frost valve has drained the system in cold weather. If such a loop is inevitable, then this pipe must be well lagged against frost.

Solar Panels below the Storage Tank. If the solar panel is partly or totally below the solar-heated tank, then the pipe

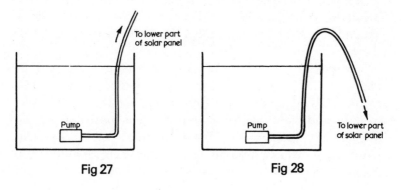

Fig 27   Fig 28

from the pump should go direct to the solar panel without rising at any point (after leaving the solar tank) so that the solar panel and pipework will drain completely when the frost valve opens (Fig 28). As explained before, anti-syphon vents in the pipes are necessary to prevent the solar water tank being syphoned out.

Remember that the anti-syphon vent (shown in the diagram) that is fixed on the delivery side of the pump must be sufficiently long to hold a column of water longer than the height to which the pump can pump it: ie longer than the working head shown on the pump's specification. Unless this is done the pump will force water out of this vent instead of through the solar panel as you require.

Anti-syphon Vents

These should be fitted as shown in Fig 21.

Insulation

Insulation of solar-heating equipment is vital to prevent loss of the valuable heat gained.

Solar-heating panels themselves should lie on a bed of insulation, as already explained, remembering that up to 40 per cent of the solar radiation can be lost via the back of the panel. A mattress made of fibreglass matting covered with polythene sheet is suitable for this purpose. A 2.5cm (1in) thickness of fibreglass enclosed in polythene will adapt itself to the contours of the solar panel. Other methods include using expanded polystyrene sheet 2.5cm (1in) thick or polyurethane sheet.

The melting point of the insulation material should be borne in mind: an empty solar panel, perhaps unused while the family is on holiday, can become very hot on a hot day and melt its insulation. For this reason fibreglass insulation is more satisfactory. Also, whereas fibreglass moulds itself to the solar panel, rigid sheeting must be cut to fit between the aluminium frames of the solar panel to present an even surface to the roof of the house. It is important not to allow wind to get under the solar panel, as it will cause additional heat loss as well as trying to rip the panel from the roof during stormy weather which could result in severe damage.

Where the solar panels are mounted on a frame the insulation must be backed with a sheet of solid material, preferably wood.

The pipes to the heated solar tank will need to be insulated, particularly in the case of flexible PVC hose as this is liable to degradation by ultraviolet rays. The pipes running from the solar system to the hot-water cylinder will also need to be insulated.

Now to the heated solar tank itself. The storage of heat here is the cornerstone of the whole operation, and great care must be taken with its insulation. It is hoped that it is situated in a loft whose roof has already been adequately insulated! Then, a close-fitting cover will be needed for the tank. It can be made from suitably treated timber. Holes must be made for the hot-water inlet from the solar panel and the outlet pipe from the pump, together with the electric cable to the pump if a submerged pump is used.

The cover is best made in two halves for ease of access.

The whole cover should be insulated with at least 10cm (4in) of fibreglass. The sides and base of the tank should also be insulated, possibly with a 10cm (4in) layer of fibreglass held in place with bands of broad sticky tape. Care must be taken that the tape does not compress the fibreglass or its insulating properties will be lessened.

A more substantial form of insulation could be made by forming a framework of hardboard and timber to surround the bottom and sides of the tank and filling it with granular vermiculite or other insulating material. A similar arrangement could be made for the lid (see Fig 29).

Pipes can be insulated by wrapping them round with strips of fibreglass or even strips of old underfelt if available. Alternatively, there are commercially made sections of expanded polystyrene pipe insulation, in sizes to fit the diameters of most pipes.

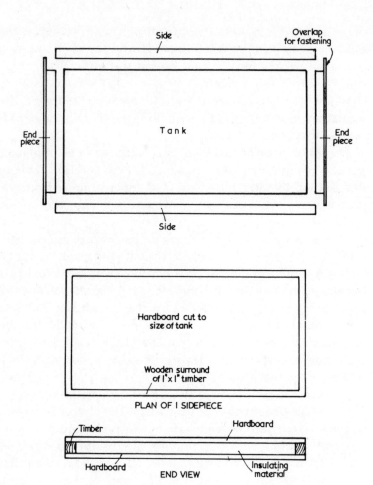

Fig 29

Electrical Connections

The electric temperature-differential controller is connected to the mains by means of a fused 13 amp plug and socket. A plug from the back of the controller conducts electricity to the pump. The leads from the two sensors (which carry only a low voltage) go to plugs in the front of the controller. These are usually colour-coded, so that the red plug goes into the red socket, etc.

One sensor is fixed in the outlet elbow of the solar panel in a position fully exposed to the sun, while the other sensor is fixed to the solar-heated tank close to the cold-water inlet. There is a knob on the front of the controller which will vary the difference in temperature between these two sensors at which the pump will operate.

Adjusting the Controller. The controller is the nerve-centre of the whole operation and controls the power supply to the pump. It sets the pump in motion when there is a difference in temperature between the sensor at the upper end of the solar panel and the sensor in the solar-heated tank.

This temperature difference can be adjusted by means of a knob on the front of the controller. The further the knob is turned anti-clockwise the smaller the temperature difference at which the pump will operate. For practical purposes this temperature difference should be about 5°C (8°F) as there must be some useful heat to be gained if the pump is to operate. Too fine a setting results in a lot of stopping and starting with minimal heat gain.

With a reasonable setting the pump will continue to operate until the temperatures equalise. Once they have done so, the pump will operate for a minute or so every ten minutes to maintain the equilibrium, which of course depends upon the sunshine available and how much water is drawn off for domestic use.

The temperatures can be checked by immersing an ordinary greenhouse-type thermometer, holding it first by the electric pump and then by the outlet from the solar panel and noting the temperature difference.

Once the controller has been set it should not be necessary to alter it further. There may be some temptation to set the control too fine in the winter but this could cause the pump to switch on when the frost valve is open. If it is cold outside (less than 10°C or 50°F) the pump may not operate even with the controller switched fully anti-clockwise. To start the pump under these conditions, either for adjustment or experiment, temporarily remove one of the small plugs from the tank sensor; it should be replaced after about one minute.

Solar-Heating in Warm Countries

In climates warmer than Britain's it may well be necessary to shade part of the solar panel to prevent the water becoming too hot.

7 Other Ways of Using Solar Energy

With a knowledge of the basic principles of solar heating you can evaluate new methods and new pieces of apparatus as they come along, or may even be able to invent your own.

One amateur in California devised a very simple system. He coiled up 300 feet of plastic hosepipe on his roof and ran cold water in at one end, displacing hot water from the other, according to his requirements. This scheme can be improved upon by adding a thermostat and a valve to regulate the flow of hot water or it can be simplified so that some water flows from it whenever the cold-water tap is turned on, provided the sun is shining.

Such a simple installation could be very useful outside a kitchen, or in a room facing anywhere near south with a suitable wall or roof receiving sunshine. It would be simple to erect this sort of solar heater, either making a separate cold-water tap for it or simply connecting the hose to the existing cold-water tap as required.

A rather more complicated scheme working on the same principles has been devised in South Africa. Here the actual solar panel is in fact its own water-storage tank.

This solar unit, as it is called, consists of a stainless-steel circular saucer-shaped piece of metal, covered with a transparent plastic dome. The whole unit is about 1 metre in diameter, stands 285mm high and is mounted on the roof, set in polypropylene insulation. Cold tap water is let into the bottom of the unit as the hot water is removed from the top. The whole unit is separate from the rest of the house water supply.

Its small absorber area (0.82 sq metre) means that it can only be used in hot countries, however. Another drawback is that when filled with water this unit is heavy, weighing 136 kilograms (2½ cwt).

Although impracticable in Britain, it does show another way in which the principles of solar heating can be adapted.

General Energy Conservation

Solar heating should not be considered in isolation. After all, it is part of the heating system of your home. Using heat from the sun goes together with saving heat in conventional ways.

For instance, experts agree that the fitting of thermostats to every centrally heated room, or even on every radiator, can produce considerable savings. The thermostats can be fixed on the actual radiators, or on the radiator but incorporating a remote sensor. Thermostats are also available for attaching to the hot-water cylinder to ensure that the water is not heated to a higher temperature than is necessary.

Insulated hot-water-cylinder jackets are now fitted almost as standard, and they save tremendous amounts of energy.

There is considerable free heat-gain in every room into which the sun shines. And there are other sources of heat gain. In a typical room incorporating a 1500-Watt radiator, half its heat will prove to be unnecessary after the first hour, due to the combined heat production of members of the family in the room, of the electric light bulbs and of the television set.

In the absence of a room thermostat the radiator continues to pump heat into the room, causing it to become uncomfortably hot and allowing considerable wastage in fuel. Fuel wastage in this way counts against any gain made by solar heating.

Next in importance comes adequate insulation in the loft, followed by insulation of the walls. Up to a third of the fuel costs in the average house is spent on heat loss through the walls. Even the best brickwork has thousands of minute gaps in it, allowing heat to escape, while wind blowing on a wet wall can cool it still further through a refrigeration effect.

Thus there is a great saving to be made by filling cavity walls with insulating material, though this needs to be expertly done. Finally, by eliminating window draughts and cutting out heat loss through the glass by fitting double glazing, you can ensure that your home is as snug and energy-saving as it is reasonably possible to make it.

Other forms of energy saving are more experimental and are not yet suitable for use in the average domestic dwelling.

To create your own electricity, wind-powered generators are noisy, potentially dangerous especially when they are large and provide electricity at only 12 volts. They are, therefore, only suitable under special circumstances and as supplementary sources. But if you are interested in making electricity for your home, read Terence McLaughlins' *Make Your Own Electricity* (David & Charles, 1977).

Water power is a possible source of electricity and suitable turbines are available for the few people who live by suitable rushing torrents. Methane-gas generators, working from poultry and other manure, are only suitable for special isolated circumstances.

8 Sources of Supply and Useful Addresses

Firms installing solar heating cannot be listed here, as so many are small concerns which may amalgamate or go out of business. But by using the information in this book you will be able to weigh up the advantages and disadvantages of any schemes that are proposed and compare one with another.

There are some useful addresses from which information is available, giving some ideas of standards and having some machinery for complaints procedures.

The National Institute of Solar Energy and Research (Epic House, Charles Street, Leicester), work with the British Standards Institute to set minimum standards by testing all the components of solar-heating systems, including pipes, panels, pumps and glass. From 1978 they have awarded a mark of approval, similar to the British Standard Kite Mark. They have a voluntary code of practice for stockists and installers, and a list of qualified stockists and installers.

The Solar Trade Association (The Building Centre, 26 Sloane Street, London WC1), have no special qualifications for membership but have a complaints procedure.

The UK Section of the International Solar Energy Society, (UK ISES, c/o Royal Institution, 21 Albermarle Street, London W1), publish literature on solar energy.

The National Centre for Alternative Technology, (Llwyngwern Quarry, Machynlleth, Powys), provides information by post on solar heating and have produced a plan for a simple solar-heating unit. They also have a permanent exhibition of solar-heating systems and other energy-conserving apparatus.

C.T.T. Ltd (Conservation Tools and Technology, 143 Maple Road, Surbiton, Surrey, KT6 48H), produce booklets on conservation including solar heating, and also sell apparatus. Their members receive a 10 per cent discount on literature, 5 per cent discount on apparatus, and also a quarterly news-sheet on natural energy.

Equipment manufacturers

Altec Solar Ltd (18 Albert Road, Richmond, Surrey), provide solar water-heating control units assembled ready for use, including fixing clamps, thermistor leads, mains cable connections and diagram. There is a choice of thermistors and one model has an indicator panel that can be mounted in the landing so that you can monitor the sensors without having to go up into the loft to see them.

Solar Save System 365 is a unit made and installed by Celestial Heating Systems Ltd (1 Crown Street, Bradford, West Yorkshire).

Thermal Systems (32 Langdale Road, Dunstable, Bedfordshire), are Suncell accredited installers. They sell a comprehensive booklet with price list, so that all the necessary equipment can be bought, plus an instruction booklet. You install it yourself or with the help of the local plumber.

Dutton-Forshaw Special Products Ltd (Moore Lane, Preston, Lancs), make solar systems working on the indirect system with a solar panel filled with oil. They also have deflecting angled concentrators to increase the efficiency and so need fewer solar panels.

Solar Economy Ltd (Balksbury Hill, Upper Clatford, Andover, Hants), design, install and offer separate components for solar water-heating systems.

Solar Heat Ltd (99 Middleton Hall Road, Kings Norton, Birmingham, B30 1AG). This company is producing all the equipment for heating domestic hot water and can supply it either in the form of a kit of parts for the DIY enthusiast or in a form suitable for local installers.

Solar Water Heaters Ltd (153 Sunbridge Road, Bradford,

West Yorkshire), manufacture solar panels.

Stellar Heat Systems Ltd (113 Stokes Croft, Bristol, BS1 3RW). For £1.75 they supply easy to follow instructions, drawings and sources of supply.

Pactrol Controls Ltd (PO Box 123, Skelmersdale, Lancs), make solar-heating controls.

Solar Traps Ltd (70 East Street, Epsom, Surrey), have a full design and installation service, suitable for any particular application of solar heating.

Spencer Solarise Ltd (Station Road, Leatherhead, Surrey), produce an indirect system using specially constituted fluid.

Index